I0487718

Lag:

A Look at Circadian

Desynchronization

Lag:

A Look at Circadian

Desynchronization

Bill Ragan, M.S.

Lulu Enterprises, Inc.

Morrisville, North Carolina

Copyright © 2007 Bill Ragan, M.S.

Published by Lulu Enterprises, Inc.

860 Aviation Parkway, Suite 300

Morrisville, North Carolina 27560

All Rights Reserved.

ISBN 978-1-4357-0221-9

Library of Congress Control Number: 2007908600

First Edition
Printed in the United States of America

Contents

Dedication

Since my earliest days, I admired the bravery of my Grandfather as he told stories of flying. Aside from his many other achievements, none appeared as appealing to me as did his experiences in aviation. I am grateful that he was able to instill in me the spirit of the aviator. This book is dedicated to him.

This book is also dedicated to many friends and teachers including George Stockert, Mike Filly, Zivko Radenkov, John Tidyman, James Henke, Chris Porr, Quince Kreb, Dr. John Wilson, Aaron Smith, Dr. John Deaton, and Dr. Leif Rivera. In addition, I would like to thank all of the pilots who are responsible for our safe passage everyday. A special thank you is also in order for those who answered my questions and forwarded me some of the information that I needed to write this book.

My dedication would not be complete if I did not also thank my wonderful wife Sara for all of her support. She has really been helpful. I also owe a thank you to my parents for all of their wisdom and emotional contributions.

Introduction

Exchrome Visual Services Photo

Introduction

At around 6:17 a.m. on August 27, 2006, Flight 5191, a Bombardier CRJ-100 crashed and burned after running off the end of Runway 26 at the Blue Grass Airport in Lexington, Kentucky. 49 of the 50 people aboard were killed in this tragic aircraft accident. At the time this was written, the National Transportation Safety Board (NTSB) had not yet released its final determinations regarding the accident involving Flight 5191 at Blue Grass Airport.

Some facts in the case of Flight 5191 remain to be discovered but, it was indicated that the aircraft was cleared for takeoff from another runway, Runway 22, when the accident occurred (NTSB, 2006). One of the many possible, yet unproven, explanations to the cause of this accident is that the pilots were suffering from the effects of sleep inertia or fatigue.

Caldwell (2002) and Mohler (1999) reported that fatigue, which can be caused by a variety of factors, including not getting enough sleep or from working for long periods of time, is dangerous. Some reasons for this are because it is difficult to self diagnose, can be enervative, and because it can cause performance deficits. These symptoms are ones that may have been present in the aviators who were piloting Flight 5191 at Blue Grass Airport as evidenced by the fact that they attempted to take off from the wrong runway. Fiorino (2006) said that this unfortunate and deadly accident occurred as a result of errors made during the taxi phase of aircraft operations. Fiorino also said that this segment of flight requires considerable concentration and awareness. This could have occurred due to any of the reasons that Caldwell and Mohler identified.

In a different accident that occurred at around 7:37 p.m. on October 19, 2004, Flight 5966 crashed short of Runway 36 on final approach at Kirksville Regional Airport in Missouri. This accident involved a British Aerospace Jetstream 32. In total, 13 people died when the aircraft collided with the ground. In this case, NTSB investigators stated in the NTSB Aviation Accident Database and Synopses that the final determination of probable cause of the accident was due, in part, to pilot fatigue (NTSB, 2006). The NTSB database's list of accidents involving pilot fatigue as probable cause in commercial and general aviation number in the hundreds since 1964, and this database excludes accidents that are only minor in nature. To reduce the possibility that an aircraft accident will occur as a result of pilot fatigue, pilots and aviation employers should take a proactive approach to preventing it from occurring. In keeping with this motive, ongoing research is being conducted in the area of pilot fatigue.

While exactly what caused the crash of Flight 5966 might not ever be known, the accident, which the NTSB attributed to pilot fatigue, may have happened because the pilots felt confused or were unable to perform as effectively as they would have if they had not been tired (Caldwell, 2002). Some might believe that it would be nice to board an aircraft as first officer, sit in the cockpit, and look over the instruments to determine the fatigue level of themselves, the copilot, and the crew before or during a flight. It is difficult, at best, to speculate when or where in the future of aviation a test like this will be feasible or permissible, but the impact that such a device might have on safety could be enormous. In the meantime, research on pilot fatigue and jet lag continues to evolve, and self-monitoring and prevention remain the state of the art for pilot fatigue management in this field.

Jet lag syndrome is a dangerous yet preventable condition that, due to various operational constraints, may occur during and after long transmeridian flights (Caldwell, 2002). This syndrome has also caused numerous difficulties for business and leisure travelers. Fatigue from traveling over long distances is often at the top of the list of travel-related complaints (Psychology Today, 1994). Rogers (1998) found that those who frequently engaged in transmeridian travel for business were more susceptible to the effects of occupational stress, more stressed, and had an increased chance of having difficulty doing their jobs well.

Various interventions have been suggested to reduce the fatiguing effects of jet lag and pilot fatigue in professional pilots and crew. Some of these interventions are sleep cycle modification, educational awareness of the signs and symptoms of jet lag syndrome and pilot fatigue, melatonin use in the form of a dietary supplement, caffeine moderation, exercise, vitamin use, and the use of sedative drugs (Caldwell,

2002). The dangerous effects of pilot fatigue in aviation, and employee fatigue in other industries, have been known to put the lives of many people in jeopardy. Caldwell also noted that this effect was further complicated by the fact that pilot fatigue was often underreported because of the stigma attached with self-reporting.

Glaus (1993) reported that the medical community has not put forth much effort into fatigue measurement, even though an accurate assessment of it could be useful when treating patients. This makes it difficult to identify pilots who are fatigued. The situation is further complicated by the fact that pilots who are fatigued may not have the skills needed to effectively cope with the additional pressures of having to work under strenuous conditions. Raymond and Moser (1995) said that lack of adequate coping skills is one problem that aviators had, and that this could result in performance problems that could result in a crash. As stated previously, there is currently no single highly effective way to objectively gauge the fatigue level of those who operate aboard aircraft.

Caldwell (2002) identified pilot fatigue as a condition that can be just as dangerous as alcohol intoxication, and that this effect can occur after as little as 18 hours without sleep. The issue of pilot fatigue was brought up by Rubin (2001), who said that it is a very real problem in aviation today, and is complicated by problems with air lines competing for landing and departure slots at busy airports. A review of current literature revealed a lack of substantial research and innovative interventions aimed at reducing the effects of pilot fatigue.

Pilot fatigue has recently become more recognized as an operational problem because of the rapid pace of business that is now commonplace (Caldwell, 2002; Hughes, 2006; Van Yperen & Hagedoorn, 2003). Rubin (2001) said that flights are now being conducted at duty hours both earlier and later than they once were to accommodate for added traffic into and out of busy airports. This problem adds to the already complicated tasks associated with piloting aircraft.

Caldwell (2002) indicated that as many as 63 million Americans already suffer from long-term daytime sleepiness that is present to a moderate or severe degree, and that 40% of all Americans reported that they did not get enough sleep each night. The particular strengths and weaknesses of each of the available fatigue interventions that are currently available to the general public are to numerous to be discussed in this book. Instead, several aspects of jet and shift lag will be discussed at length here. In particular, consideration for educational awareness of the signs and symptoms of jet lag syndrome will be mentioned. In addition, interventions that are now available to reduce lag's enervative effects, especially ones that may be related to occupational stress in pilots will be a point of focus in the next several chapters.

CHAPTER 1: Behind the Lag

U.S. Air Force Photo

Hardaway and Gregory (2005) said that pilot fatigue has resulted in many aviation accidents over recent years and that these accidents can obviously be costly and have a broad range of negative consequences and effects. Among other reasons, jet lag as a result of long sector lengths is to blame for fatigue. Rubin (2001) said that it is the passengers who stand to lose the most when pilots suffer from the effects of fatigue, and that they may be unknowing victims of the performance deficits that may result from pilot fatigue. Further complicating the job for professional pilots, occupational stress was shown by Henn (1995) to exist in air line transport pilots because of pilot fatigue.

While pilot fatigue has been studied and has been identified as one problem for aviators, the relationship between stress and pilot fatigue intervention training and occupational stress has not been established. Training in aviation has been repeatedly shown to be an effective way to cope with mechanical difficulties when they arise, as Hytten (1989) found. Hytten said that in a review of one case where a helicopter crashed in water, all but one of the five crewmembers survived. The one who did not survive had not received any training on how to egress from the aircraft in the event of such a landing. Hunter (2005) also indicated that training was an effective intervention when discouraging pilots from risk-taking attitudes.

There are several reasons to be interested in the underlying performance reducing causes of pilot fatigue and jet lag syndrome. Some of these reasons have been frequently identified as a cause of circadian disruption (Hardaway & Gregory, 2005). Not the least of those reasons is to help to prevent disastrous crashes, such as those that may result from cognitive impairment. There are several ways that the effects of lag can be reduced. One way includes employer motivation, as Hunter, Martinussen, and Wiggins (2003) put it. Hunter et al. said that one way to make flying safer is to encourage employers to be more forgiving about policies regarding aviators calling off of duty. He said that pilots who call off instead of report to duty when overly fatigued may be part of a safer fleet than those pilots who have employers who are not forgiving of pilots who call off. The effects of fatigue on cognition should not be understated, and one source of fatigue that occurs in aviation results from jet lag.

Though age and gender have been identified as variables that could have an effect on how pilots cope with occupational stress, the areas of interest in this book are sector length and, separately, pilot fatigue prevention training. Jet lag syndrome can result in uncomfortable and potentially dangerous effects to pilots, passengers, and crewmembers. These conditions can extend to problems with health and performance. This situation may be exacerbated by high operational requirements and lack of educational and preventative programs to address the specifics of the problems that aviators may face. It is unknown whether or not training and education about jet lag syndrome may be beneficial at reducing occupational stress and pilot fatigue.

CHAPTER 2: The History of Lag

U.S. Air Force Photo

William C. Ocker, the father of "Blind Flying," flew from Brooks Field, San Antonio, Texas, to March Field, California in a covered cockpit to prove that a pilot's instrumentation could be more reliable than his own natural senses.

P ast and present research on fatigue in aviation and intervention strategies to reduce the effects of fatigue during aviation operations will be discussed in detail in this chapter. The research was conducted mainly to attempt to examine possible factors that influence accidents that have occurred in air carrier operations, also known in the United States as Part 121 operations (FAA, 2006). While conducting research related to this subject, some historical problems that have occurred in the past were also considered, and will be discussed in the following chapters.

Reducing Fatigue in Aviation

Pilot fatigue as a result of jet lag, high-tempo operations, and side effects from over-the-counter medications has been established by several researchers as being one reason why otherwise safe flying and aircraft maintenance operations can become unsafe. This is particularly true when sleep cannot be achieved before duty, or when it is interrupted because of precluded sleep acquisition (Aarons, 2003; Caldwell, 2002; Mohler, 1999; Northrup, 1997; Waterhouse &

Reilly, 1997). All forms of aviation operations are susceptible to the effects of pilot fatigue. Luckily, research in this area has expanded from earlier days. In days gone by, noise and altitude sickness were believed to be the cause of fatigue. This attitude has transitioned to recent attitudes and research that has been focused on various biological changes that occur along with long distance travel and lag. One interesting finding is that there have been reports that the size of different parts of the brain may have changed after prolonged exposure to jet lag. Researchers have indicated that this atrophy of the brain may have occurred because of prolonged stress that has resulted from the continued circadian desynchronization caused by recurring cases of jet lag syndrome (Cho, 2001). While these findings are certainly obscure, they are also very applicable. How can such outcomes be stopped or reversed?

Caldwell and Brown (2003) said that combat and contingency operations in military aviation can be complicated by pilot fatigue, and that this can even occur when factors like poor weather, long missions, and difficult terrain simply make duties more difficult to perform. These added stressors, in some cases, along with other factors, can compound the problems aviators face. This confluence of factors can result in airplane crashes and the subsequent death of crewmembers (Fiorino 2004; Fiorino, 2006; Mohler, 1999).

Since different people are affected by fatigue in different ways, as Waterhouse et al. (2000) indicated, it can be difficult to accurately predict who will suffer. Also, it is difficult to predetermine what symptoms they will have, or when they will have them. Individuals who have been affected by acute desynchronization have been observed with physical detriments that span several biological systems. Another scientific observation that has been made was related to body temperature. It was found that regular patterns of bodily temperature and sleep rhythms, mood, and performance have been interrupted as a result of jet and shift lag.

While systems have been shown to be disrupted by lag, several factors have been shown to reduce the acute effects of lag, as well. These factors include providing prior training and experience with jet lag and one's genetic differences that might have an influence on how jet lag is experienced from one person to the next (Caldwell, 2002; Monk, Moline, & Graeber, 1988; Petrie, Powell, & Broadbent, 2004; Waterhouse & Reilly, 1997).

Recent advances in the sciences have permitted more detailed imaging of the brain. With these remarkable advances, more accurate recording of brain rhythms has become possible. With enhanced imaging techniques, the chronic effect of desynchronization as a result of fatigue and jet lag has emerged as a new topic of scientific interest. Several studies have linked chronic desynchronization to health problems. In each of two different studies, Cho (2001), and Waterhouse et al. (2000) both found that frequent recurring circadian desynchronization resulted in physiological and psychological problems including lack of awareness, fatigue, and poor cognition. Along with circadian desynchronization, increased stress and other health problems were also observed. Of particular interest here is the possibility that circadian desynchronization causes occupational stress.

The underlying reasons for performance deficits in pilots will be developed more fully throughout this chapter, but of some importance is the effect of the hormone cortisol. This hormone was also found to be produced during periods of stress. The hormone cortisol has also been linked to the stressful effects of fatigue, which has been shown to result from desynchronization. While some researchers have found that there was an interaction between cortisol and stress, Kurina, Schneider, and Waite (2004) did not find a significant correlation between the two. Even though the link between cortisol and stress was not well substantiated in these two studies, more research in this aspect of circadian desynchronization, in the true experimental study form, seems to be needed.

Since there are varying reports on the relationship between cortisol levels and stress level, more research into this area is needed. Daurat, Benoit, and Buquet (2000) said that the length of sleep and the quality of sleep were negatively affected by circadian desynchronization. Both of these detriments have also been determined by Daurat et al. to have been some possible underlying causes of stress.

Even when pilots were unable to identify the effects of fatigue in themselves, researchers found that underlying hormonal variation was occurring. In a study conducted in 2004 by Samel, Vejvoda, and Maass, pilots who conducted sustained operations and did not have any overt physiological signs of fatigue still had higher salivary cortisol levels, possibly indicating the presence of an undetected, underlying stressor.

Fatigue has been implicated in research articles as a problem in aviation for almost 100 years. It has been named more often in recent years, as transmeridian flights have become more common over the past 50 years. But it was not until more recently that the discussions changed from being focused on environmental factors to being focused on circadian desynchronization as the cause of pilot fatigue. Even more recently, the recurring effects of acute desynchronization began emerging as a topic of study.

Past Research

Pilot fatigue and attempts to measure it have been an important topic of consideration by scientists because of the effects that lack of awareness have had on safety. The reasons behind the acute and chronic effects of fatigue have perplexed scientists and researchers, and further developments in this area have been continuing to occur. In days when aviation science was still in the early stages of development, Feree and Rand (1937) suggested that planes and pilots both needed a screening before flight.

In an early attempt to achieve this, Feree and Rand (1937) recommended that a tachistoscope be used to determine the fitness of pilots just prior to flight. Feree and Rand said that fatigue level was related to accommodation speed in the eye. This theory may have been quite true: LeDuc, Greig, and Dumond (2005) conducted a study and showed that fatigue and ocular response times were indeed related.

Over 20 years after the first transatlantic flights were beginning to occur around 1919, McFarland (1941) conducted research which included participants who made transoceanic flights. McFarland found that pilot fatigue was the result of poor fitness levels of the pilots, lower oxygen content at altitude, poor diet, and alcohol and tobacco use. McFarland also investigated human factors in engineering as an aspect of fatigue, and cited noise, temperature, and vibration as problems.

In another effort to identify problems in the cockpit, Randall and Patt (1947) noted that the pilot's workspace could be better optimized for comfort in an effort to reduce pilot fatigue. McFarland (1943) also found that diet and smoking were among the top problems in aviation medicine. Other problems included aeroembolism and body temperature variation as a result of operation at high altitude. Trends in research articles from this era in aviation also showed that scientists were seeking a way to reduce cockpit fatigue while at the same time they were trying to explain the nature of it.

As clinicians and researchers continued to identify and attempt to remedy some of the general problems in the growing field of professional aviation in the 1940's and 1950's, the problems of fatigue and exposure were among the most frequently studied. At the same time, aviation was expanding quickly. There became a sudden need for more pilots as well, as Day, Miller, White, and Baldwin (1944) indicated. These authors found that as the need for pilots grew, the regular standards by which they were selected were more and more frequently overlooked. This caused many pilots who would normally be disqualified for flight to be placed on duty rosters.

Around this same time, scientists also tested the possible use of prescription drugs to overcome some of the physiological and psychological difficulties that pilots were experiencing. In 1947, Davis considered the administration of amphetamines as a possible way to counter the fatigue that aircrews were feeling after prolonged duty. There were problems noted with the drug interventions, though.

Davis (1947) indicated that amphetamines were not a very good solution to the fatigue problem. Davis said that stress and anxiety were the underlying reasons that pilots were having problems, and that it was these problems, not long duty days, that were to blame. Also, Davis then said that amphetamines were effective at waking pilots from fatigue, but did not increase their accuracy level while performing duty tasks.

Since accuracy was not increased, the benefit of drug interventions to reduce fatigue seemed limited in efficacy. The idea of using drugs to reduce the effects of fatigue in aviation has been frequently discussed, and the topic will emerge again under other headings in this chapter. Also, instead of discussing all of the fatigue interventions further here, the discussion will now turn to the early use of technology and other innovations that were employed in the fight against fatigue.

In 1943, Clinton and Thorn studied pilot fatigue in senior aviators, paying special attention to physical and mental symptoms that could arise from duty. These authors used an electroencephalogram (EEG) machine as part of the study, and found 80% of the pilots that were involved in the study had normal EEG readings. Clinton and Thorn found that those who they tested suffered from fatigue from self reported problems with irregular scheduling and lack of quality lodging between trips. A question remained, though. What could be done about it?

In this particular case, the technology behind the EEG was beneficial at providing extra insight into a problem in aviation. Other technology used in the advancement of the aviation industry has also been shown to actually cause fatigue: Russell, Erwin, and De Haven (1943) said that the increased range of aircraft, along with increased speed and noise in the cockpit as a result of emerging technology were all factors that actually contributed to pilot fatigue. On the topic of pilot fatigue, Russell et al. said that this was a problem that could be overcome through proper selection and training of pilots.

Russell et al. (1943) also indicated that the ability to withstand fatigue varied from one pilot to the next due to their individual differences. These findings were simultaneously emerging during the same time period that the development of the turbojet engine was being completed. Flaiz and Wolf's (1943) findings that cockpit fatigue could be linked to excessive noise, particularly during take off and landing configurations, were also made at about the time in history, and Van Der Linden (1992) and Winter and Van Der Linden (1992) said that it was during this period that the jet age in America began.

In the 1940's, as the jet age was developing and the world was at war, rapid changes in aviation technology took place (Goetz, 2006). Alongside the technological changes, the scientific consideration of the mental and physical health of pilots continued, as Wright (1947) noted. Following stress and pilot fatigue studies, Wright found that some aviators who had undergone extremely difficult combat missions exhibited symptoms of acute anxiety including sleep loss, tremors, and phobic reactions.

Wright (1947) indicated that the symptoms were actually a survival response that did not hinder, but rather furthered their ability to fly. Wright suggested that clinicians should not be fast to judge such reactions as a neurosis when they are useful skills for piloting. Wright did not fully differentiate between the side effects of vigilance and the symptoms of fatigue, of which the former has been shown to be an antecedent to the latter.

In 1967, well after the end of the Second World War, Hauty studied various aspects of human physiology including rectal temperature, heart rate, reaction time, and decision time both before and after transmeridian flights. Hauty found that only during east-west flight was performance significantly impaired, and that the effects took several days to subside. The article by Hauty is particularly noteworthy because it appears to have preceded others in regards to desynchronization due to transmeridian flight.

Since the article by Hauty (1967), jet lag has been the subject of several scientific experiments and literature reviews. In one study, Murakami and Murakami (1980) tested the possibility that jet lag syndrome could be reduced or eliminated in mice in the laboratory by causing them to become acclimatized to a new time zone ahead of time.

Contrary to Murakami and Murakami's (1980) belief that acclimatization was a solution to jet lag syndrome, the central control system of the mice they used in the experiment did not respond to the training. These researchers indicated that the regulatory system within the body is more resilient to change than they originally believed it would be.

In the time following Murakami and Murakami's (1980) study, the researchers conducting studies on jet lag syndrome slowly began to implicate the pineal gland in the brain as being responsible for the production of melatonin. In addition to being a key proponent of sleep, this gland was also found to be responsible for controlling sleep-wake cycles and mood (Sprinivasan,

1989). Along with other discoveries in fatigue symptomology and prevention, the discussion of using drugs to train the body to new time zones was frequently cited in the research that was reviewed.

Authors at Science News (1975) wrote that certain drugs could be used to switch the body clock forward or backwards. They also indicated that research in that day was pointing to theophylline, a stimulant, and phenobarbital, a depressant, as the drugs that might have some control over the body's ability to do it. The Science News authors said that that theophylline could cause a phase forward shift of 18 hours. Of the drug phenobarbital, they said that it could be used to set the human clock back 12 hours. Turek and Van Reeth (1998) also discussed the use of medicine to retrain the body clock. They said that drugs like the bezodiazepine triazolam could also be used to phase shift the circadian cycle forward.

Turek and Van Reeth (1989) further emphasized the possibilities of using triazolam to help counter jet lag syndrome, and said that the drug caused significant changes to both the behavioral and endocrine circadian rhythms. Another drug, synthesized melatonin, was found by Wright, Aldhous, Franey, English, and Arendt (1986) to be successful at inducing evening rest, and did so without changing the mood of the participants. Wright et al. noted that melatonin did not cause any undesirable effects on the endocrine system of the participants in a double-blind cross-over test.

Cassone (1990) emphasized the importance of the natural hormonal regulation of body rhythms, and said that synthesized melatonin as a supplement was useful in treating circadian desynchronization. There has been a controversy surrounding discussion about the use of synthesized melatonin to reduce jet lag, and several different opinions have emerged.

By the 1980's, there was an abundance of literature on jet lag. In an article from this era in aviation, Tapp and Holloway (1981) said that they performed a series of laboratory experiments and concluded that retrograde amnesia occurred after circadian desynchronization. Tapp and Holloway's tests came at a time in history when researchers were also actively studying the effects of travel on human performance in business and sports environments. Loat and Rhodes (1989) further substantiated the dangers of the desynchronization that occurs during transmeridian flight when they identified jet lag syndrome as having deleterious effects on both the physiological and the psychological cycles in athletes.

Loat and Rhodes (1989) also said that during periods of desynchronization, body temperature, metabolism, hormonal excretion, arousal, rest cycles, and reaction time were all negatively affected. The authors said, as others also have, that these effects varied from person to person, and could be affected by age, fitness level, and extent of desynchronization. These findings seem to help to substantiate the value and importance of a well implemented stress and fatigue reduction program.

In another effort to further explain the possible deficits that jet lag may cause, Tapp and Natelson (1989) performed tests with rhesus monkeys that demonstrated that after a small phase shift of just six hours, both vigilance and discrimination were impaired. Tapp and Natelson said that vigilance was more strongly affected than discrimination, and that in vigilance, the negative effect lasted longer. The authors of this study also indicated that a second performance deficit that was observed ten days to two weeks after the initial six hour phase shift occurred.

In a study on ways to reduce the effects of desynchronization, Bower (1987) said that people who are more active suffer less from desynchronization than those who are not. After performing a laboratory study, Bower found that hamsters recovered from induced circadian desynchronization after running on a treadmill and that they did so much faster than hamsters that did not exercise. This reduction of symptoms may be especially important when considering that even a small improvement in this area may result in a significant reduction of losses that might occur because of aircraft accidents as a result of fatigue.

Monk (1989) was also interested in forecasting fatigue level prior to flight, and tested the use of the Visual Analog Scale (VAS) to help predict decreasing energy levels as they occur. Monk found that Global Vigor (GV) assessed with the VAS was found to be sensitive to jet lag syndrome.

Buck, Tobler, and Borbély (1989) were also interested in monitoring fatigue level of aviators, though they considered a different, more technologically advanced approach than Monk (1989). Buck et al. used a method of monitoring bedtime motor activity by having participants wear a wrist device after long flights. The authors said that the use of such a device was an effective way of measuring the quality of sleep that pilots received after flying for extended durations.

The biological basis of acute jet lag syndrome may have the most important clues to the explanation and prevention of this potentially hazardous condition. Brown (1990) said that pineal activity is much more active at night than it is at day, and that this activity comes when the gland secretes melatonin, its primary hormone. When this secretion occurs, the melatonin adjusts and helps to regulate the body's natural clock. Brown said that this occurrence is the sole result of the pineal gland's production of melatonin, which is predominantly synthesized from the pineal gland.

Because the synthesis of melatonin is triggered by low light conditions, night flights can be particularly dangerous when the effects of increased melatonin output and high levels of fatigue are combined.

Courtney (1994) said that in one study of five crewmembers making a transoceanic flight during the early morning hours, all of the crewmembers displayed brain waves that were the same as those found in controls taken when they were either asleep, or very sleepy. Courtney said that the crewmember levels in this study were recorded between the hours of 0400 and 0600, and that the aircraft was on autopilot at the time.

Some of the early problems with the design of aircraft that caused problems for pilots, like operation at altitude with limited cabin pressure and in cold temperatures have mostly been improved or eliminated. These changes may have, to some degree, made air operations easier on pilots, but other issues that have caused problems for pilots have since been discovered. Various changes in the aviation industry may have caused still other difficulties for pilots to have to cope with.

Not the least of the emerging problems is the current high tempo of operations and the use of competitive business models that have been the cause of low cost fares. These are just some of the things that may have caused difficulties where some others were previously relieved.

CHAPTER 3:
The Evolution of Jet Lag

NASA Photo

In 2003, Caldwell and Brown said that going without sleep for one night can result in an approximate loss of 25% to 30% of psychological ability, an approximate 60% loss after two days without sleep, and an approximate 80% loss after four consecutive days without sleep. Jet lag syndrome has been shown to directly affect the ability to sleep in the hours soon after desynchronization occurs. As Fiorino (2001) indicated, there are cases where acute pilot fatigue from duty and acute jet lag syndrome cause a conflicting condition where sleep is not achieved when it is time to rest. Along with acute jet lag syndrome, the chronic effects of desynchronization have also been under increasing investigation by researchers recently.

In one study on the chronic effects of desynchronization, Cho (2001) observed flight attendants who were repeatedly exposed to jet lag, and found that significantly more atrophy was present in the right temporal structures of the participants who commonly had a short recovery time of 5 days after flights across 7 time zones when compared to those who had 14 days of recovery time after similar flights.

Cho's (2001) research in this area will be discussed more thoroughly later in this chapter, but the implications from the research that Cho conducted could suggest a need to bring changes to the number of consecutive years that a pilot may perform long-haul flights in a given career if the health risk revealed by these researchers is more widely considered.

In another study, Caldwell, Hall, and Erickson (2002) conducted research with UH-60 helicopter pilots to investigate the possibility that EEG information could be consistently recorded and translated during flight. Their efforts were a successful attempt to determine if the pilots were experiencing increased fatigue levels while at the aircraft controls. Though highly advanced, the use of EEG in scientific studies is not new, and Caldwell et al. affirmed the validity of the EEG by saying that using the EEG is the best way to gain insight into the ongoing workings of the central nervous system (CNS).

This information, when paired with earlier findings, helps to indicate that certain EEG activity levels can be used to effectively evaluate sleep deprivation. Research prior to that conducted by

Caldwell et al. (2002) also reliably linked EEG levels, fatigue, and pilot performance levels, specifically in slow-wave brain activity. The early research in this area had been conducted in laboratories, but not in-flight.

When Caldwell et al. (2002) took EEG readings during flight, they broke new ground in aviation psychology. They found that, like had previously only been tested and confirmed in the laboratory, slow-wave brain activity, and specifically theta wave activity, was increased among sleep deprived pilots who underwent in-flight EEG testing. The authors of this study determined that it is possible to interpret the fatigue levels of pilots while in-flight by theta wave activity as indicated by EEG readings.

Dussault, Jouanin, and Guezennec (2004) also explored the use of the EEG and electrocardiography (EEC) equipment as a means to determine fitness level of aviators and found that delta and theta activity was increased during high workload flight sequences, and that the alpha frequency levels were reduced. These authors also found that the reverse was true during times of rest.

Dussault et al. (2004) also indicated that an average increase in heart rate of 8.89 beats per minute (bpm) was observed from active flight phases to periods of rest, with the increase in bpm during flight. In a study by Mace and Carroll (1989), even with the employment of stress control interventions, the rise of heart rate in response to stress was not reduced. This was seen even though the skill level of the group who received the intervention was higher than the control group. Dobkin and Pihl (1992) observed heart rate increase during stressful conditions and said that the stressful conditions were found to be the antecedent to the rise of heart rate.

The use of the EEG method of determining fatigue level was more effective and reliable than the Traveler Profile Questionnaire that was developed in 2003 by Flower, Irvine, and Folkard. Flower et al. indicated that the Travel Profile Questionnaire, a tool that they created to rate the extent of travel related fatigue as self-reported by the participant, did not give a reliable indication of the effects of fatigue. Aside from the use of the EEG, the use of futuristic methods of intervention for pilot fatigue was only discussed in the most recent literature.

One such device, the Electronic Pilot-Activity Monitor (EPAM), was developed by Cabon, Bourgeois-Bougrine, Mollard, Coblentz, and Speyer (2003). According to its creators, this device was attached to the wrist of pilots while in flight, and used to monitor alertness levels or time naps that were taken while in flight. While the authors noted that the EPAM was not fully tested or determined fail-safe at the time that the article was written, they did indicate that current testing showed that the device was useful, to a degree, at detecting pilot drowsiness while flying. Hummel (2000) indicated that the approved use of napping while on duty (one of the many possible ways to reduce fatigue during flight), and the EPAM may help to better facilitate napping as a means to ameliorate the effects of pilot drowsiness.

The review of literature revealed that research articles on acute jet lag syndrome often include the discussion of ways to prevent the active symptoms of it. A study by Petrie, Powell, and Broadbent (2004) that focused on methods of primary prevention found that 13% of all the commercial pilots they surveyed indicated that they were significantly fatigued from their job as an aviator, and that daytime napping was an effective way to reduce their fatigue.

During the course of their study, Petrie et al. (2004) also discovered that 19% of the crew that they surveyed used a prescribed medicine to help them sleep, or melatonin or some other form of non-prescribed sleeping agent in the two month period before the study. These findings help to substantiate the need for discovering other ways that pilot fatigue interferes with job tasks. In another study by Deixelberger-Fritz, Tischler, and Kallus (2003), it was found that pilots who participated in a fatigue inducing exercise in a laboratory performed markedly better on a choice reaction time task and a performance in concentration test when given an energy drink. The energy drink they gave participants was made with caffeine 80mg, taurine 1000mg, and glucuronolactone 600mg additives. This was given to the participants before hand, while another group of participants received the same energy drink but without the additives.

As a result of their experiment, Deixelberger-Fritz et al. (2003) recommended that pilots use breaks, and breaks with nutrition to combat the effects of fatigue that are encountered while performing their duties. Whether interventions like these are often used by pilots to combat fatigue or if their employers actively encourage pilots by frequently offering stress and fatigue reduction training programs is not known but may be the subject of future investigation.

Another group of scientists questioned the use of caffeine and modafinil to reduce the effects of fatigue, believing that they may have a reducing effect on g tolerance in pilots who took them. Putting their beliefs to the test in an experiment with adult male rhesus monkeys, Florence et al. (2005) found that modafinil and caffeine had no significant effect on gravity loss of consciousness (G-LOC) because of reduced cerebral blood flow when compared to controls. While this research may not have much influence on transport operations, these findings may be important for those who partake in military operations.

A time that G-LOC may factor into transport operations is during emergency procedures, in which case knowing about a possible interaction could be important. Based on findings from their experiment, Florence et al. (2005) said further testing should be conducted with human participants to confirm the results that they obtained.

Recent research has shown that in the case where destination time zones are different from home time zones, or where duties must be performed overnight, the question of how to achieve optimal sleep during daytime hours is occasionally raised. Caldwell, et al. (2003) conducted an experiment to determine if the effects of the hypnotic drug temazepam would be useful in promoting better sleep during the daytime.

The experimenters found that those who received the temazepam slept better than those who received a placebo, and also found that those who received the drug had better results on the Psychomotor Vigilance Task than those who had been administered a placebo. Where permitted, drug interventions, along with other approaches, can help to decrease occupational stress in aviators. This is done by helping to control some of the stress that can go along with the job.

Commercial pilots are not the only ones to use medicine and napping to counter the effects of fatigue. Kenagy, Bird, Webber, and Fischer (2004) reported that one group of military pilots who were engaged in long combat missions used the drug dextroamphetamine on 58% of their missions and in-flight napping on 94% of them. On shorter missions, the same group reported using dextroamphetamine on 97% of their missions and napping on 13% of them. Caldwell (2002) said that dextroamphetamine, also called Dexedrine, or Dex for short, has been found to be effective at reducing fatigue in laboratory and real-world tests.

Caldwell (2002) warned readers that Dex is no substitute for adequate rest, and said that those who used the drug were able to maintain normal cockpit vigilance while using it and experienced no significant physiological or psychological side effects. Caldwell also noted that for several hours after the dose of Dex wore off, pilots experienced mild difficulty achieving normal sleep. The risks and advantages associated with drug use to combat fatigue or to aid sleep during desynchronization is discussed later in this chapter.

Using a chi-square test to investigate the correlation between proportions of accidents to length of duty exposure, Goode (2003) found that there was an increased probability of accidents with longer duty times. This also raises some suspicion that occupational stress levels will be higher when length of duty is prolonged. Goode suggested that those who determine how many hours a pilot will be performing their duties should consider that there may be an increased safety risk when a pilot is scheduled for a longer period of duty.

Bourgeois-Bougrine, Cabon, Mollard, Coblentz, and Speyer (2003) indicated that during short-haul flights, morning flights represented work that was more plagued by fatigue and increased workload than afternoon flights because of the amount of sleep that pilots tended to achieve before each one. In another study, Conway, Mode, Berman, Martin, and Hill (2005) found that in a case-control analysis where pilots were grouped according to accident rates in an almost 11 year period, case pilots had fewer career flight hours than control pilots, yet the case pilots worked more hours per week than the control pilots. Why was this the case?

The case operators were less likely to take into account pilot duty time than the control operators were, and the case pilots were much more prone to fly into unknown weather conditions than the control pilots. Conway et al. (2005) said that the combination of these factors made accidents more likely in the case group than in the control group and that the pilots who had less flight hours worked more and crashed more than those with more flight hours.

Current Perspectives on Jet Lag

Gander et al. (1998) found that among one group of crewmembers who made long-haul trips, the average number of time zones crossed in one 24 hour period was 8. Among the same crewmembers, it was found that they slept for 105 minutes less on trip layover sleep periods than they did prior to the trip. During the four or five days at a time when the crewmembers were on duty, it was observed that they were more tired, drank more caffeinated beverages, snacked more but ate full meals less, and suffered from more minor aches and pains and sinus troubles than they did during days prior to duty.

From this research, it appears as though long-haul pilots, in general, may be under more occupational stress than short-haul pilots for several reasons, including that they average less sleep than they are accustomed to getting, and that their work environment requires them to adapt to a schedule that promotes desynchronization. Since this hypothesis has not been tested in any of the research that was reviewed, it may become an important topic for researchers to consider. This effect may be further exacerbated by the increase in the tempo of operations in commercial aviation that has been reported as occurring today.

All of the preceding conditions are believed to be caused, according to Bennett (2003), in part by the operational requirements needed to sustain budget air line model of operations that is frequently being used today. Veillette (2005) said that fatigue is as much an issue today as ever, and that pilots who are fatigued are more prone to make mistakes, and that while fatigued, there is a greater possibility that small control errors will be made. Research by Haynes (2003) showed that depression and sleep difficulties were found to exist together, but it was not indicated which condition was present first. Changes to the biological clock as a result of transmeridian flight can cause various difficulties for pilots.

Akerstedt, Kecklund, Gillberg, Lowden, and Axelsson (2000) said that fatigue hits those affected by desynchronization the worst on the first day of trying to reacclimatize, and that pilots who fly across many time zones suffer from fatigue on their days off more so than do their counterparts who fly short-haul trips. Whether the fatigue on days off that Akerstedt et al. discussed was present in long-haul pilots was the direct result of acute jet lag syndrome or of chronic stress as a result of prolonged long-haul flying was not something that they discussed, but both may be a factor.

Biological variation can cause different degrees of symptoms for those who are experiencing the symptoms of jet lag. With this in mind, another factor to consider is that Veillette (2005) indicated that pilots who were over 45 years old reported that they have more problems sleeping and sleep fewer hours than younger pilots did. Add to this the effect of stress on the job from operating in a post-9/11/2001 environment and one where operational tempo is high (Bennett, 2003), and one may not be surprised that some pilots complain of fatigue and stress.

Extreme or debilitating, fatigue can be dangerous or fatal, and Beiswenger (2002) noted that sleep approaches quickly, and sometimes without warning, and worse yet, that it can happen to anyone. The modern perspective that jet lag and fatigue in aviation is a dangerous component of travel today is also emphasized by the current Federal Aviation Administration (FAA) regulations. These rules stipulate that each pilot must only accrue 8 hours of flight time in a 24-hour period, and only then if the pilot has gotten 8 consecutive hours of sleep in the preceding 24-hour period (FAA, 2006; MacDonald, 2005).

FAA (2006) authors indicated that regulations regarding the number of hours that pilots can perform duties have been in effect since the 1940's. There is controversy surrounding the more recent rulings, according to Richfield (2001), who said that the Air

Line Pilots Association (ALPA) did not believe that the FAA guidelines protect against pilot fatigue enough.

CHAPTER 4:
Conditions Marked by Fatigue

N ot all of the conditions listed in this chapter will happen to all people all of the time; but reports of them have been made. All of the conditions listed will have various environmental and genetic influences. Research on some of them is still being conducted.

Musculoskeletal pain.

In a survey conducted by Simpson and Porter (2003), a correlation between total flight experience, annual flight hours, and lower back pain existed. Simpson and Porter said older pilots are more prone to suffer from back pain than younger pilots, and that musculoskeletal pain had little effect on a pilot's actual flight performance. Whether or not the pain is caused solely by environmental factors or may have been directly linked to jet lag syndrome was not discussed. Similar findings were also reported by Gander et al. (1998) in a study where self reports and observations of long-haul pilots were used to determine the effects of acute jet lag syndrome.

Gander et al (1998) stated that jet lag syndrome was not to blame for the reported pain, though, because in another study of short-haul pilots by the same authors, similar back pains were reported by those participants as well. Gander et al. said that possible causes of the back discomfort reported by the participants were lack of adequate preflight rest and the effects of alcohol, caffeine, and nicotine use by crewmembers.

Haugli, Skogstad, and Hellesøy (1994) also surveyed both long- and short-haul pilots and found that low back pain was among the most frequently reported of all complaints by both categories of pilots.

Temporal lobe atrophy.

Cho (2001) found that increased cortisol activity occurred with longer flights, and that the result of that activity was atrophy of the temporal lobe which occurred after approximately five years of such recurrent activity. Cho stated that amnesia has also been shown to occur in individuals with damage to the hippocampus, and that individuals who were employed by an air line for over five years and had short recovery periods of less than five days between flights across at least seven time zones had higher salivary cortisol levels than a group who had longer recovery periods.

Bosch (2000) had similar findings, and indicated that long-haul flying, increased salivary cortisol levels, and memory impairment were related. Though no additional research was available in the area of temporal atrophy as a result of jet lag, endocrine disorders present in studies in general medicine that resulted from high cortisol levels were well documented (Elgh et al., 2006; MacLullich et al., 2005; Tarquis, 2006).

Findings of higher cortisol activity resulting from stress, and stress as a result of sleep deprivation was also well documented (Bosch, 2000; Cho, 2001; Copinschi, 2005; Samal et al., 2004).

Performance deficits.

Considering that it can take more than one day for each time zone crossed to recover from the debilitating effects of jet lag, it is not surprising that research indicated that a decline in performance resulted from circadian desynchronization (Armstrong, 2006; Drust, Waterhouse, Atkinson, Edwards, & Reilly, 2005; Reilly, Waterhouse, & Edwards, 2005). Courtney (1994) said that a 75 % decline in the ability to perform complex cognitive tasks occurred in those who do not have experience dealing with jet lag syndrome, and a 10 %

decline occurred even in those who have been specifically trained to deal with the debilitating effects of fatigue.

Drust et al. (2005) said that experts in the field of chronobiology have determined that certain sports activities were performed better at certain times of the day than others, and that disruption of regular circadian rhythms had a negative effect on performance.

Stress.

Bosch (2000) found that flight attendants who traveled across more than eight time zones per week had higher levels of output of the stress hormone cortisol when traveling across several time zones, but not when their trips were not transmeridian. Bosch said that problems sleeping and problems with short term and memory processing have been linked to increased cortisol output. Cortisol, a hormone secreted by the adrenal gland and has been linked with arousal, has also been shown to have negative effects on memory and cognition (Bosch, 2000; Cho, 2000; Copinschi, 2005; Samal et al., 2004).

Van Yperen and Hagedoorn (2003) also indicated that high demand jobs cause employees to become exhausted, and that there is a subsequent need for rest. Van Yperen and Hagedoorn said that when the employee has a high degree of control in their job, whether it is a highly demanding position or not, that they will have lower occupational stress.

Cancer.

Filipski, Li, and Levi (2006) performed laboratory tests to determine if jet lag and malignant tumor growth were related, and found that some indicators of increased malignant growth were present in tumor models present in mice that had their rhythms altered. In an earlier study, Filipski, et al. (2004) performed a case study on women in aviation who had breast cancer, and found that family history was the leading risk factor, but the authors also linked desynchronization with the breast cancer. Any link between recurrent desynchronization and cancer needs to be more fully investigated, but it will not be done so here.

In 2003, Bourgeois-Bougrine, Carbon, Gounelle, Mollard, and Coblentz conducted a survey where pilots reported that their source of fatigue was due to lack of adequate sleep as a result of duty scheduling, late night flying, and jet lag. The individuals who suffered from fatigue where generally not very good at assessing their own level of performance, but methods of self examination and examination by others have been shown to exist, and can be used to measure fatigue (Veillette, 2005). Efforts that have been made to improve deficiencies in this area may have helped to reduce the risk of unsafe flight that might have occurred as a result of fatigue, especially when the risk for fatigue was at its greatest.

Gander, Rosekind, and Gregory (1998) said that during long-haul flights, the chance for pilots to be required to perform flight duties through periods of circadian lows was increased, and with that came higher risk that fatigue would strike while duties were being performed. When pilots were asked to perform duties under such conditions, the risk of suffering from debilitating fatigue was greater. The suggested use of an EPAM device was just one way that the risks associated with fatigue while flying could have been limited and that might have had lifesaving potential (Bourgeois-Bougrine et al., 2003).

Reilly and Bambaeichi (2003) also advocated the use of wrist mounted actimetric devices that routinely recorded activity levels based on the time of day in which activity occurred after transitioning from one time zone to another as a means of assessment for scientific purposes. Another method of fatigue assessment was noted in 2005 by LeDuc et al., who found that a correlation between fatigue and ocular responses after flight in the AH-64 Apache helicopter existed. The authors said that both pupil size and constriction latency were greater and that constriction amplitude and saccadic velocity was lesser after flight than before. Pilot participants also said that they were more tired after the flight than before the flight.

This type of testing may be further developed in the days to come, and an idea similar to the one that Feree and Rand (1937) had might be regularly implemented to test pilots prior to flight, or in an on-the-spot manner if fatigue is suspected. Another assessment tool by Lamond, Dawson, and Roach (2005) was shown to be useful for measuring fatigue. These authors found that the psychomotor vigilance task (PVT), of which a 5 and 10 minute version is available, was accurate in identifying both sleep loss and fatigue. How this tool can be used to make reduce the risk of accidents that result from fatigue is not yet known. In the next chapter, interventions will be discussed. They are important because they can help us to reduce the risk of injury from fatigue related accidents.

CHAPTER 5:

Interventions for Jet Lag

B ecause of the many different negative outcomes that have come from cockpit fatigue, several interventions have been developed to reduce its dangerous effects (Driskell & Mullen, 2005). Finding a better method to enable aviators to withstand the effects of pilot fatigue may be one way to make flying safer. In an effort to do this, Hardaway and Gregory (2005) found that offering training to pilots on how to cope with fatigue before a long-haul mission, along with offering them the opportunity to take an extra layover day to rest was effective at improving their duty performance and reducing their reported fatigue levels.

Primary prevention of jet lag is one strategy that employers of aviation professionals can take to attempt to increase the performance of pilots. There are numerous ways to reduce the effects of acute jet lag syndrome, some of which are described in the following sections.

Primary Prevention

Hummel (2000) said that in-flight napping was one way to reduce fatigue in the cockpit, and that taking short naps while the aircraft was piloted by another crewmember helped to reduce the effect of fatigue and actually improved performance during phases of flight that required the heightened vigilance of all required crewmembers. Napping is a form of intervention that falls into both the primary prevention category and into the employer sponsored prevention category because in-flight napping should be conducted as part of standardized cockpit procedures, and may be best conducted with the help of a wrist worn device to prevent excess napping (Wright, Powell, McGowen, Broadbent, & Loft, 2005).

Researchers found that one risk with napping was the effect of sleep inertia, which could affect performance after waking from a nap (Hofer-Tinguely et al., 2005). Veillette (2005) said that pilots reported that they received about one hour less sleep than usual before the first day they return to duty, and each layover night when on duty.

The effects of in-flight napping were also investigated by Driskell and Mullen (2005), who said that increased safety was one possible outcome when an in-flight napping program is properly implemented.

In another effort towards primary prevention, Charlton (2005) defined a system of ongoing personal awareness of self-diagnosis, treatment, and monitoring (S-DTM) that might apply in the case of crewmember fatigue. Charlton said that the more often the S-DTM model is used, the more accurate the individual performing it will be at it, and that the treatment element can be either a therapeutic intervention or a drug treatment.

The principle behind this model is that the aviator is aware of the onset of their own fatigue, counteracts it with good health habits, diet, and exercise, and then monitors how they progress. Courtney (1994) said that preflight fatigue was the best predictor for the onset of fatigue during flight. This fits into the self-diagnosis stage of the S-DTM model, because pilots who are able to identify their own fatigue level accurately may also be able to take early interventions against fatigue.

Courtney (1994) offered several suggestions that may help reduce the risk of becoming fatigued while on duty, including getting frequent exercise, maintaining continued motivation to work and perform well on the job, having a healthy diet, and drinking adequate water.

The S-DTM model may already be the one that pilots are using but are not aware this is the case. Bourgeois-Bougrine et al. (2003) said that attempts to get more, and better rest was the method that pilots surveyed stated that they used the most to overcome fatigue.

Overall good health habits have been successfully used as a way to reduce the effects of stress and fatigue. Giannakoulas (2003) found this to be true in one study, and said that when pilots who smoke are occasionally required to not smoke, such as may occur during a prolonged flight, they often complain of fatigue and difficulty concentrating, among other things.

Since getting enough rest in the first place is the primary intervention of choice, it seems appropriate to attempt to find ways to promote this behavior. Outside of using medicine to facilitate sleep, Geraghty (2005) said that having a regular and repeatable routine before sleep may help to quickly trigger bed time rest. This strategy should be carefully planned though, because habits may be easily and frequently broken by those who travel away from home.

According to Geraghty (2005), by using the same routine before bed time, both while at home and while traveling, it may be possible to become trained to quickly fall asleep soon after a period of performing crew duties.

Employer Sponsored Prevention

One reason why corporate fatigue interventions have been important in the prevention of accidents is because of the number of people that have been rapidly affected by the implementation of a single program tends to be large. Hummel (2000) recommended the NASA Ames Fatigue Countermeasure Program as a way to introduce good rest habits to pilots who may be at risk for suffering from fatigue. This type of program can be viewed by all members of a department at one time, ensuring both the consistency and reach of the intervention.

In another approach to fatigue prevention, Veillette (2005) said that it is not so much the number of hours that the pilot flies, but the timing of the flying that can make a difference in how pilots feel when performing their duties. By implementing a prevention plan that teaches pilots how to be aware of and how to reduce the active symptoms of jet lag syndrome and by advocating the use of careful scheduling strategies by dispatchers, the risk for pilot fatigue related incidents may be further reduced.

This suggestion came at a time when a complicated method of scheduling that took into account the personal circadian rhythm of the individual pilot that was being scheduled for duty was being considered by the Aviation Rulemaking Committee for future FAA Part 125/135 rules (Jackson, 2005). Another way to reduce pilot fatigue risks was suggested by Gamauf (2004), who said that a specialized computer training program was available from a company called *Word One* that may have been able to help reduce or prevent fatigue in aviation, and that met Canada's CAR 573.16 regarding required training.

Driskell and Mullen (2005) concurred, saying that employers could help pilots by carefully constructing work schedules that interfere with their sleep schedules as little as possible. A carefully planned strategy used by those who create duty rosters for pilots is one way that this could be realized.

On the topic of human factors in aviation, Grant (2002) suggested that the comfort level of the cockpit should be carefully considered during its design because such errors can result in fatigue causing musculoskeletal stress and other problems. Smith (2006) also found that complaints of fatigue were related to sustained exposure to noise and vibration in aircraft.

In another area of human factors in aviation, cabin air quality has also been linked to fatigue, and it is possible that employers could take extra care to ensure that the most advanced methods of filtering the air were in place to reduce the risk of pilot fatigue as a result of exposure to dirty, stale air (Lindgren & Norbáck, 2005).

The Authorized Use of Drugs

Caldwell and Caldwell (2005) indicated that personal behavioral and employer sponsored interventions should be the first line of defense against jet lag syndrome, and said that drug interventions are sometimes required as well. The military has authorized several such fatigue fighting drugs, and Caldwell and Caldwell said that caffeine, dextroamphetamine, and modafinil can all be effective in situations where such use is both permitted and required. In one study, Caldwell, Caldwell, and

Darlington (2003) found that dextroamphetamine could be effective in military aviation operations as a fatigue deterrent for pilots.

Among the other drugs that have gained approval in specialized settings such as military aviation, Caldwell, Caldwell, Smith, and Brown (2004) specifically identified modafinil as a drug that could be used to increase vigilance during flight. The authors warned that the drug should be used with caution, as its use did not bring performance up to the same level as it would be at if adequate rest had been obtained prior to flight. The risk of overuse due to the marginal effects of this drug was also indicated. Caldwell and Caldwell (2005) recommend that pilots seek alternatives to the use of drugs to counteract fatigue before using them.

As with the many other changes that have occurred in technology since the dawn of aviation, so have theories of how to perform better while operating aircraft changed. The operation hazards that caused pilot fatigue in aviators during aviation's early years have been significantly reduced, but only to bring rise to other difficulties that came about as long-range travel was conducted faster and became more common. Fatigue that was once believed to be the sole cause of uncomfortable cockpit and cabin environments has recently been attributed to operational tempo and biological tendencies that may result from eastward travel across several time zones.

Though it has been known that transmeridian flight and high cortisol levels were linked, at the time of this writing it was not known to what extent if any occupational stress stays present in aircrew that made frequent transmeridian flights. It was also undetermined how findings that indicated that recurrent desynchronization and cancer were linked would affect the overall long term health of those who partook in such operations.

Several interventions were discussed to reduce the effects of jet lag, and one prevailing attitude is health psychology related. Under this model, the pilot would be fit to fly, be drug and alcohol free, and be willing to commit to getting rest when needed to be best prepared for upcoming flights. Courtney (1994) said that prior prevention is the best intervention for fatigue, and the literature reviewed does not seem to present any better solution to the problem of fatigue.

Because separate literature that was reviewed indicated that stress and fatigue is a preventable condition that could cause accidents in aviation, attempts to identify the various causes of stress and fatigue were pursued. During this endeavor, literature revealed many elements that can and do contribute to occupational stress in pilots, but no research was found to have been conducted testing occupational stress among pilots using a reliable measure.

Research revealed no previous efforts to determine if a correlation exists between proactive pilot stress and fatigue intervention training and occupational stress. Also, no efforts to link any of the various causes of stress and fatigue, including jet lag and pilot fatigue to occupational stress were found to have been conducted.

CHAPTER 6:
Summary

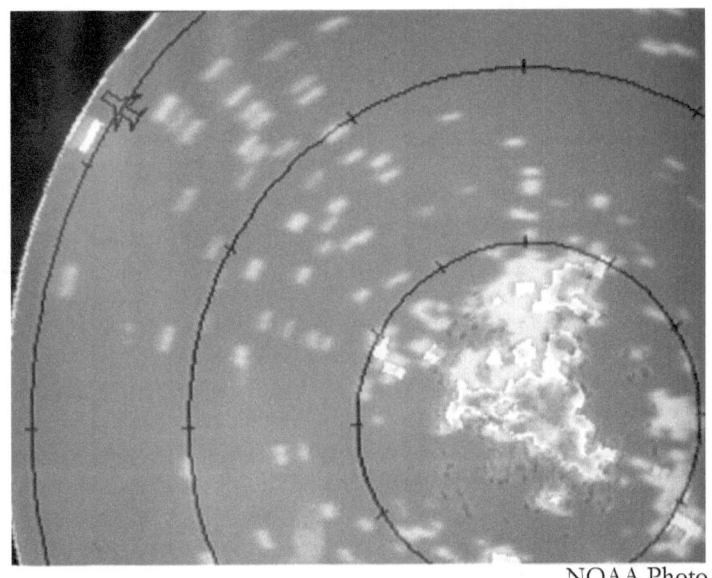

NOAA Photo

The discussion of fatigue and circadian desynchronization that has preceded this chapter has been limited to the field of aviation, but research in this area has been and can be applied to other industries as well. Fields of work that require 24-hour operational coverage like public safety, emergency medicine, and some areas of the manufacturing industry have been prone to have employees who routinely suffer from the symptoms of fatigue (Van Yperen & Hagedoorn, 2003). In these fields, it is not jet lag that occurs, but a similar condition, shift lag, does.

Often times, when an individual is required to work day shifts for some period of time, and then switches to night shifts for a period of time, or works night shifts exclusively, this problem occurs. The problem of desynchronization occurs in the first case very frequently, and can happen in the second case when an individual has to cope with trying to get quality sleep during daytime hours.

Critical Review of Literature

Much has been written about jet lag and circadian desynchronization over the past few decades. From the early days of aviation all the way through to today, researchers, psychologists, medical doctors, and pilots have been searching for remedies to some of the problems that come naturally when our body clock is disrupted or when operating in an environment that is different than the one that we are commonly accustomed to.

Fatigue from jet lag can be dangerous (Aarons, 2003; Caldwell, 2002; Mohler, 1999; Northrup, 1997; Waterhouse & Reilly, 1997). To make matters worse, everyone is susceptible to the effects of fatigue, and the resultant stress has been linked to airplane crashes (Fiorino, 2004; Fiorino, 2006; Mohler, 1999). Each person who experiences the combined physical and psychological effects that can result from jet lag, stress, and pilot fatigue does so differently and with different intensity (Caldwell, 2002; Monk, Moline & Graeber, 1998; Petrie et al., 2004; Waterhouse & Reilly, 1997). Samel et al. (2004) said that one thing that makes assessment of fatigue complicated is that its symptoms are often difficult to recognize, even by the individual who is affected by it.

In the 1940's, the FAA began to regulate crew rest requirements (FAA, 2006). This marked the first formal recognition of crew rest as factor that could result in safer flight. During the same era, scientists were working to overcome some of the many causes of crew fatigue through innovation and research (Feree & Rand, 1937). One of them, MacFarland (1941) was looking into complaints of fatigue made by pilots and found that noise, temperature, and low oxygen content at altitude were all partly to blame for the way pilots felt after flight. Another innovation in the 1940's was the use of drugs to delay fatigue (Davis, 1947). As flights became longer and aircraft became faster, the research continued.

As transmeridian and transoceanic flights became more commonplace, research into fatigue was ongoing. Eventually, a link between desynchronization and fatigue was made (Haughty, 1967). With jet technology and high tempo operations, fatigue management is still a focus of research and development, and drug and non-drug interventions for fatigue and jet lag are still being developed and perfected (Caldwell, 2003; Cassone, 1990; Kenagy, Bird, Webber, & Fischer, 2004; Murakami & Murakami, 1980; Science News, 1975; Turek & Van Reeth, 1988).

Other Points

Some problems in aviation, such as mechanical reliability, efficiency, and quality of cabin accommodations have been effectively solved over the past several decades. Even though technological advances have provided a level of relief to aviators, some other technological advances have actually presented difficulties to pilots. In the case of jet lag today, pilots do not so much have to directly deal with many of the discomforting aspects of flight that were one time normal with operations at altitude. These problems included lack of breathing air and exposure to cold temperatures, but they now are faced with needing split second reflexes and to cope with desynchronization as passenger and military jets are much faster now than in the past.

Pilots are also faced with the pressure of having to operate in a difficult business environment where certain management models may require them to function with maximum efficiency (Van Yperen & Hagedoorn, 2003).

Social Implications

As the world has become more globally connected over the past few decades by transportation, shipping, and communications systems, certain logistical problems have emerged. Just as other companies have obligations to their customers, those in aviation do too (Phillips, 2006). The theory of corporate social responsibility (CSR), according to Oketch (2005), extends across a wide variety of areas but always takes into account the way in which a business conducts itself. When applying the principals of CSR, the aviation corporation would always consider the customer, the environment, and the employees before considering profits. To this end, corporate employers in aviation would seek to invest in new strategies and technologies to improve CSR.

This is occurring as an overall business trend, according to Herro and Tazzara (2006). Herro and Tazzara said that socially responsible investing (SRI) is a growing trend among Fortune 500 companies, and that more companies will be employing this strategy in the future. Employing CSR techniques may also help to attract qualified new employees, according to Ray

(2006). Ray said that new job applicants distinguished between companies with CSR programs and without them.

Brief Review

Fatigue, as a result of jet lag and circadian desynchronization is one potential risk that emerges in part due to modern operational requirements. With this risk comes the need for further exploration into and explanation of its causes. Fiorino (2001) said that one of the problems with pilot fatigue is that the technology that is being used today is better than the humans who operate it, and that programs designed to teach pilots to manage stress and fatigue are part of the next chapter in aviation. Fiorino also said that some of these programs were already being developed by some companies, and that other ways of promoting stress and fatigue prevention should be actively encouraged.

Fiorino (2006) also said that the NTSB recently recommended that additional training on stress and fatigue prevention be provided by the FAA in order to lessen the chance that dangerous aviation accidents will occur as a result of these conditions. Since there are many things that can reduce pilot performance and increase pilot fatigue, including jet lag, unusual duty hours, extended duty periods, foul weather, cockpit comfort level, and lack of adequate rest, there are many areas that pilot fatigue prevention can target in order to improve conditions in this area (Mohler, 1999).

It may be that prevention programs are one of the best ways to reduce the risk of the dangerous effects of pilot fatigue, because attempts in the past to make changes to pilot duty days has been full of controversy. Croft (2001) said that there remains a difference in what ALPA and the FAA believe is adequate time to rest. Increased rest may be one way to solve the pilot fatigue problem, but several interventions can be applied to help maximize what time pilots currently have between flights.

This can be particularly important when it is not the number of hours on duty that causes the fatigue, but the type of duty performed when on duty that is the issue, as Samel, Wegmann, and Vejvoda (1997) indicated. In research by these authors, it was found that pilots complained of debilitating fatigue levels during certain flights because of time of day, length of flight, and flight conditions, even when the duty times were less than the FAA maximums.

For organizations that do not already have a stress and fatigue prevention program in place, they may find some benefit in implementing one, according to Hardy (1992). Hardy found that performance enhancing strategies, including positive self-talk and stress reduction techniques, could be applied to enhance overall resiliency and hardiness. While research showed that prevention programs are beneficial, no scientific data were found to exist to demonstrate that a quantifiable correlation exists between prevention programs and a well documented predictor of performance, occupational stress.

The problem of lag is rooted in a genetic response to conditions that go back to the age of cave men. It was believed that humans rooted down in the woods at night to avoid predators and because their vision was not good enough to keep them from getting hurt while hunting after dark. This situation, along with a biological need for rest to conserve energy may have developed into circadian rhythm as we experience it today. Because the imprints are so well established, it is unlikely that any one quick fix will resolve lag as a result of transmeridian flight or working the night shift.

Many easy to implement interventions appear to be available, though, to allow those who experience these stresses on the job to have some relief. If some of the interventions named in Chapter 5 are too difficult to implement, try starting with one or two that are easy for you to use. Also, be sure to be aware of your own fatigue level, as many researchers have said that this is still the best indicator of fatigue that is available today.

References

Aarons, R. N. (2003). Fatigue: the subtle killer.
   Business & Commercial Aviation, 93(5), 100-104.

Akerstedt, T., Kecklund, G., Gillberg, M., Lowden,
   A., & Axelsson, J. (2000). Sleepiness and days of
   recovery. *Transportation Research Part F: Traffic
   Psychology and Behavior, 3*(4), 251-261.

Armstrong, L. E. (2006). Nutritional strategies for
   football: Counteracting heat, cold, high altitude,
   and jet lag. *Journal of Sports Sciences, 24*(7), 723-740.

Beiswenger, R. (2002). Fatigue: Insidious tool of the
   grim reaper. *Flying Safety, 58*(6), 10-12.

Bennett, S. A. (2003). Flight crew stress and fatigue
   in low-cost commercial air operations – an
   appraisal. *International Journal of Risk Assessment,
   4*(2/3), 207-231.

Bosch, X. (2000). Stress hormone impairs long-term
   retrieval of memorized information. *Lancet,
   355*(9209), 1078.

REFERENCES

Bourgeois-Bougrine, S., Carbon, P., Gounelle, C., Mollard, R., & Coblentz, A. (2003). Perceived fatigue for short- and long-haul flights: A survey of 739 air line pilots. *Aviation, Space, and Environmental Medicine, 74*(10), 1072-1077.

Bourgeois-Bougrine, S., Cabon, P., Mollard, R., Coblentz, A., & Speyer, J. (2003). Fatigue in aircrew from short-haul flights in civil aviation: The effects of work schedules. *Human Factors and Aerospace Safety, 3*(2), 177-187.

Bower, B. (1987). Hamster jet lag: Running it off. *Science News, 132*(23), 358.

Buck, A., Tobler, I., & Borbély, A. A. (1989). Wrist activity monitoring in air crew members: A method for analyzing sleep quality following transmeridian and north-south flights. *Journal of Biological Rhythms, 4*(1) 93-105.

Brown, G. M. (1990). Chronopharmacological actions of the pineal gland. *Drug Metabolism and Interactions, 8*(3-4), 189-201.

Cabon, P., Bourgeois-Bougrine, S., Mollard, R., Coblentz, A., & Speyer, J. (2003). Electric pilot-activity monitor: A countermeasure against fatigue on long-haul flights. *Aviation, Science, and Environmental Medicine, 74*(6), 679-682.

Caldwell, J. A. (2002). Fatigue factors for aviators…And everybody else! *Flying Safety, 58*(10), 20.

Caldwell, J. A., & Brown, L. (2003). Runnin' on empty? *Flying Safety, 59*(3), 4.

Caldwell, J. A., & Caldwell, J. L. (2005). Fatigue in military aviation: an overview of US military-approved pharmacological countermeasures. *Aviation, Space, and Environmental Medicine, 76*(7), 39-51.

Caldwell, J. A., Caldwell, J. L., & Darlington, K. K. (2003). Utility of dextroamphetamine for attenuating the impact of sleep deprivation in pilots. *Aviation, Space, and Environmental Medicine, 74*(11), 1125-1134.

Caldwell, J. A., Caldwell, J. L., Smith, J. K., & Brown, D. L. (2004). Modafinil's effects on simulator performance and mood in pilots during 37h without sleep. *Aviation, Space, and Environmental Medicine, 75*(9), 777-784.

REFERENCES

Caldwell, J. A., Hall, K. K., & Erickson, B. S. (2002). EEG data collected from helicopter pilots in flight are sufficiently sensitive to detect increased fatigue from sleep deprivation. *International Journal of Aviation Psychology, 12*(1), 19-32.

Caldwell, J. L., Prazinko, B. F., Rowe, T., Norman, D., Hall, K. K., & Caldwell, J. A. (2003). Improving daytime sleep with temazepam as a countermeasure for shift lag. *Aviation, Space, and Environmental Medicine, 74*(2), 153-163.

Cassone, V. M. (1990). Effects of melatonin on vertebrate circadian systems. *Trends in Neurosciences, 13*(11), 457-464.

Charlton, B. G. (2005). Self-management of psychiatric symptoms using over-the-counter (OTC) psychopharmacology: The S-DTM therapeutic model--Self-diagnosis, self-treatment, self-monitoring. *Medical Hypothesis, 65*(5), 823-828.

Cho, K. (2001). Chronic 'jet lag' produces temporal lobe atrophy and spatial cognitive deficits. *Nature Neuroscience, 4*(6), 567-568.

REFERENCES

Clinton, M., & Thorn, G. W. (1943). Studies on commercial air line pilots. *War Medicine, 4*, 363-373.

Conway, G. A., Mode, N. A., Berman, M. D., Martin, S., & Hill, A. (2005). Flight safety in Alaska: Comparing attitudes and practices of high- and low-risk air carriers. *Aviation, Space, and Environmental Medicine, 76*(1), 52-57.

Copinschi, G. (2005). Metabolic and endocrine effects of sleep deprivation. *Essential Psychopharmacology, 6*(6), 341-347.

Courtney, S. D. (1994). Fatigue management – New insight. *Flying Safety, 50*(8), 2-5.

Croft, J. (2001). Air lines, pilot unions split on crew rest. *Aviation Week & Space Technology, 154*(22), 38.

Daurat, A., Benoit, O., & Buquet, A. (2000). Effects of zopiclone on the rest/activity rhythm after a westward flight across five time zones. *Psychopharmacology, 149*(3), 241-246.

Davis, D. (1947). Psychomotor effects of analeptics and their relation to fatigue phenomena in air-crew. *British Medical Journal, 5*, 43-45.

REFERENCES

Day, E., Miller, R. B., White, L., & Baldwin, J. M. (1944). Medical problems in an overseas transport service. *Journal of Aviation Medicine, 15*, 2-8.

Deixelberger-Fritz, D., Tischler, M. A., & Kallus, K. W. (2003). Changes in performance, mood state, and workload due to energy drinks in pilots. *International Journal of Applied Aviation Studies, 3*(2), 195-205.

Dobkin, P. L., & Pihl, R. O. (1992). Measurement of psychological and heart rate reactivity to stress in the real world. *Psychotherapy and Psychosomatics, 58*(3-4), 208-214.

Driskell, J. E., & Mullen, B. (2005). The efficiency of naps as a fatigue counter-measure: a meta analytic integration. *Human Factors, 47*(2), 360-377.

Drust, B., Waterhouse, J., Atkinson, G., Edwards, B., & Reilly, T. (2005). Circadian rhythms in sports performance – an update. *Chronobiology International, 22*(1), 21-44.

Dussault, C., Jouanin, J., & Guezennec, C. (2004). EEG and ECG changes during selected flight sequences. *Aviation, Space, and Environmental Medicine, 75*(10), 889-897.

Elgh, E., Lindqvist, A. A., Fagerlund, M., Erikkson, S., Olsson, T., & Nassman, B. (2006). Cognitive dysfunction, hippocampal atrophy, and glucocorticoid feedback in Alzheimer's disease. *Biological Psychiatry, 59*(2), 155-161.

Federal Aviation Administration, (2006). *Pilot flight time and rest.* Retrieved September 26, 2006, from http://www.faa.gov/news/fact_sheets/news_sto ry.cfm?newsId=6762.

Feree, C. E., & Rand, G. (1937). Human factor in airplane crashes. *Archives of Ophthalmology, 18*, 789-795.

Filipski E., Delaunay, F., King V. M., Wu, M. W., Claustrat, B., Gréchez-Cassiau, A., et al. (2004). Effects of chronic jet lag on tumor progression in mice. *Cancer Research, 64*(21), 7879-85.

Filipski, E., Li, X. M., & Levi, F. (2006). Disruption of circadian coordination and malignant growth. *Cancer Causes and Controls, 17*(4), 509-514.

Fiorino, F., (2001). Wake-up call for fatigued pilots. *Aviation Week & Space Technology, 155*(3), 82-84.

REFERENCES

Fiorino, F., (2004). FedEx crash factors. *Aviation Week & Space Technology, 160*(24), 38.

Fiorino, F., (2006). Pilot fatigue rules assessed. *Aviation Week & Space Technology, 164*(5), 44.

Fiorino, F., (2006). Fateful turn. *Aviation Week & Space Technology, 165*(9), 107.

Flaiz, J. W., & Wolf, E. G. (1943). The effect of noise on aviation personnel. *Contact, 3*, 160-167.

Florence, G., Riondet, L., Serra, A, Etienne, X, Huart, B., Van Beers, P, et al. (2005). Psychostimulants and G tolerance in Rhesus monkeys: effects of oral modafinil and injected caffeine. *Aviation, Space, and Environmental Medicine, 76*(2), 121-126.

Flower, D. J. C., Irvine, D., & Folkard, S. (2003). Perception and predictability of travel fatigue after long-haul flights: a retrospective study. *Aviation, Space, and Environmental Medicine, 74*(2), 173-179.

Gamauf, M. (2004). Fatigue risk management course. *Business and Commercial Aviation, 95*(4), 142.

Gander, P. H., Gregory, K. B., Miller, D. L., Graeber, R. C., Connell, L. J., & Rosekind, M. R.

(1998). Flight crew fatigue II: Short-haul fixed-wing air transport operations. *Aviation, Space, and Environmental Medicine, 69*(9), 8-15.

Gander, P. H., Gregory, K. B., Miller, D. L., Graeber, R. C., Connell, L. J., & Rosekind, M. R. (1998). Flight crew fatigue V: Long-haul air transport operations. *Aviation, Space, and Environmental Medicine, 69*(9), 37-48.

Gander, P. H., Rosekind, M. R., & Gregory, K. B. (1998). Flight crew fatigue VI: A synthesis. *Aviation, Space, and Environmental Medicine, 69*(9), 849-60.

Geraghty, L. N. (2005), Paradise lost, *Health, 19*(6), 184-187.

Giannakoulas, G. (2003). Acute effects of nicotine withdrawal syndrome in pilots during flight. *Aviation, Space, and Environmental Medicine, 74*(3), 247-51.

Glaus A. (1993). Assessment of fatigue in cancer and non-cancer patients and in healthy individuals. *Supportive Care In Cancer: Official Journal Of The Multinational Association Of Supportive Care In Cancer, 1*(6), 305-15.

REFERENCES

Goetz, T. J. (2006). Birth of the American jet age. *Aviation History, 17*(1), 22-29.

Goode, J. H. (2003). Are pilots at risk of accidents due to fatigue? *Journal of Safety Research, 34*(3), 309-313.

Grant, K. A. (2002). Ergonomic assessment of a helicopter crew seat: The HH-60G flight engineer position. *Aviation, Space, and Environmental Science, 73*(9), 913-918.

Hardaway, C. A.; Gregory, K. B. (2005). Fatigue and sleep debt in an operational Navy squadron. *International Journal of Aviation Psychology, 15*(2), 157-171.

Hardy, L. (1996). Testing the predictions of the Cusp Catastrophe Model of anxiety and performance. *The Sport Psychologist,* 10(2), 140-156.

Haughty, G. T. (1967). Individual differences in phase shifts of the human circadian system, and performance deficit. *Life Sciences and Space Research,* 5, 135-147.

Haugli, L., Skogstad, A., & Hellesøy, O. H. (1994). Health, sleep, and mood perceptions reported by air line crews flying short and long hauls.

Aviation, Space, and Environmental Medicine, 65(1), 27-34.

Haynes, P. L. (2003). *Circadian impact of psychosocial factors in depression.* Unpublished doctoral dissertation, University of California, San Diego.

Henn, S. M. (1995). *Occupational and psychosocial stress among airline transport pilots: behind the power curve.* Unpublished doctoral dissertation, Pacific Lutheran University, Washington.

Herro, A., & Tazzara, C. (2006). Capitalism grows more socially conscious. *World Watch, 19*(6), 8.

Hofer-Tinguely, G., Acherman, P., Landolt, H. P., Regel, S. J., Rétey, J. V., Dürr, R., Borbüly, A. A., et al., (2005). Sleep inertia: performance changes after sleep, rest, and active waking. *Cognitive Brain Research, 22*(3), 323-331.

Hughes, D. (2006). Stormy weather. *Aviation Week & Space Technology, 164*(13), 46-47.

Hummel, T. (2000). The long haul. *Flying Safety, 56*(6), 18-22.

Hunter, D. R. (2005). Measurement of hazardous attitudes among pilots. *International Journal of Aviation Psychology, 15*(11), 23-43.

REFERENCES

Hunter, D. R., Martinussen, M., & Wiggins, M. (2003). Understanding how pilots make weather-related decisions. *International Journal of Aviation Psychology, 13*(1), 73-87.

Hytten, K. (1989). Helicopter crash in water: effects of simulator escape training. *Acta Psychiatrica Scandinavica: Supplementum, 355,* 73-78.

Jackson, K. S. (2005). Testy rest and duty issues. *Business and Commercial Aviation, 96*(4), 88.

Kenagy, D. N., Bird, C. T., Webber, C. M., & Fischer, J. R. (2004). Dextroamphetamine use during B-2 combat missions. *Aviation, Space, and Environmental Medicine, 75*(5), 381-386.

Kurina, L. M., Schneider, B., & Waite, L. J. (2004). Stress, symptoms of depression and anxiety, and cortisol patterns in working patients. *Stress & Health: Journal of the International Journal for the Investigation of Stress, 20*(2), 53-63.

Lamond, N., Dawson, D., & Roach, G. D. (2005). Fatigue assessment in the field: validation of a hand held electronic psychomotor vigilance task. *Aviation, Space, and Environmental Medicine, 76*(5), 486-489.

REFERENCES

LeDuc, P. A., Greig, J. L., & Dumond, S. L. (2005). Involuntary eye responses as a measure of fatigue in U.S. Army Apache Aviators. *Aviation, Space, and Environmental Medicine, 76*(7), 86-91.

Lindgren, T., & Norbáck, D. (2005). Health and perception of cabin air quality among Swedish commercial air line crew. *Indoor Air, 10*(15), 65-72.

Loat, C. E., & Rhoades, E. C. (1989). Jet-lag and human performance. *Sports Medicine, 8*(4), 226-238.

MacDonald, E. (2005). Only pilots get tired and other urban myths. *Air Medical Journal, 24*(2), 63-65.

MacLullich, A. M. J., Deary, I. J., Starr, J. M., Ferguson, K. J., Wardlaw, J. M., & Seckl, J. R. (2005). Plasma cortisol levels, brain volumes and cognition in healthy elderly men. *Psychoneuroendocrinology, 30*(5), 505-515.

Mace, R. D., & Carroll, D. (1989). The effects of stress inoculation training on self-reported stress, observer's rating of stress, heart rate, and

REFERENCES

gymnastics performance. *Journal of Sports Sciences,*
7(3), 257-266.

McFarland, R. A. (1941). Fatigue in aircraft pilots.
New England Journal of Medicine, 225, 845-855.

McFarland, R. A. (1943). Some problems in aviation
medicine. *Bulletin of the New England Medical Center,*
5, 1-6.

Mohler, S. R. (1999). Pilot fatigue manageable, but
remains insidious threat. *Flying Safety, 55*(5), 16.

Monk, T. H. (1989). A Visual Analogue Scale
technique to measure global vigor and affect.
Psychiatry Research, 27(1), 89-99.

Monk, T. H., Moline, M. L., & Graeber, R. C. (1988).
Introducing jet lag in the laboratory: Patterns of
adjustment to an acute shift in routine. *Aviation,*
Space, and Environmental Medicine, 59(8), 703-710.

Murakami, H., & Murakami, Y. (1980). Effect of an
altered rest-activity or feeding schedule on the
shift of motor activity rhythm of mice. Aviation,
Space, and *Environmental Medicine, 51,* 371-374.

National Transportation Safety Board. (2006).
Aviation Accident Database & Synopses. Retrieved
October 6, 2006 from http://www.nstb.gov.

REFERENCES

Northrup, S. E. (1997). Melatonin and aircrew. *Flying Safety, 53*(10), 16-17.

Oketch, M. O. (2005). The corporate stake in social cohesion. *PJE. Peabody Journal of Education, 80*(4), 30-52.

Petrie, K. J., Powell, D., & Broadbent, E. (2004). Fatigue self-management strategies and reported fatigue in international pilots. *Ergonomics, 47*(5), 461-468.

Phillips, E. D. (2006). Corporate social responsibility in aviation. *Journal of Air Transportation, 11*(1), 65-87.

Psychology Today (1994). A clockwork cocktail. *Psychology Today, 27*(6), 8-9.

Ray, J. R. (2006). Investigating relationships between corporate social responsibility orientation and employer attractiveness. Unpublished doctoral dissertation, The George Washington University, District of Columbia.

Raymond, M. W., & Moser, R. (1995). Aviators at risk. *Aviation, Space, and Environmental Medicine, 66*(1), 35-39.

REFERENCES

Randall, F. E., & Patt, D. I. (1947). The physical principles involved in pilot comfort and efficiency. *Journal of Aviation Medicine, 18*, 184-191.

Reilly T., & Bambaeichi, E. (2003). Methodological issues in studies of rhythms in human performance. *Biological Rhythm Research, 34*(4), 321-336.

Reilly, T., Waterhouse, J., & Edwards, B. (2005). Jet lag and air travel: implications for performance. *Clinics in Sports Medicine, 24*(2), 367-380.

Richfield, P. (2001). FAA sets a 16-hour duty time limit. *Aviation Week & Space Technology, 88*(2), 19.

Rogers, H. L. (1998). *A survey of the travel health experiences of international business travelers.* Unpublished doctoral dissertation, University of Calgary, Canada.

Rubin, R. (2001). It's time to update rest and duty-time rules. *Aviation Week & Space Technology, 154*(22), 66.

Russell, W. E., Erwin, J. R., & De Haven, H. R. (1943). Medical research in some aspects of aircraft design. *Journal of Aeronautical Science, 10*, 227-231.

REFERENCES

Samal, A., Vejvoda, M., & Maass, H. (2004). Sleep deficit and stress hormones in helicopter pilots on 7-day duty for emergency medical services. *Aviation, Space, and Environmental Medicine, 75*(11), 935-940.

Science News. (1975). Undoing future shock: A jet lag pill. *Science News, 108*(2), 25.

Simpson, P. A., & Porter, J. M. (2003). Flight-related musculoskeletal pain and discomfort in general aviation pilots from the United Kingdom and Ireland. *International Journal of Aviation Psychology, 13*(3), 301-318.

Smith, S. D. (2006). Seat vibration in military propeller aircraft: characterization, exposure assessment, and mitigation. *Aviation, Space, and Environmental Medicine, 77*(1), 32-40.

Sprinivasan, V. (1989). The pineal gland: its physiological and pharmacological role. *Indian Journal of Physiology and Pharmacology, 33*(4), 263-72.

Staier, M. (2005). The night shift: Can we completely adapt and how? *Approach: The Naval Safety Center's Aviation Magazine, 50*(3), 22-25.

REFERENCES

Tapp, W. N., & Holloway, F. A. (1981). Phase shifting circadian rhythms produces retrograde amnesia. *Science, 211*(4486), 1056-1058.

Tapp, W. N., & Natelson, B. H. (1989). Circadian rhythms and patterns of performance before and after simulated jet lag. *The American Journal of Physiology, 257*(4), 796-803.

Tarquis, N. (2006). Hypothèses neurobiologiques concernant les liens entre psychopathie et maltraitance infantile. *L'encéphale, 32*(1), 377-84.

Turek, F. W., & Van Reeth, O. (1988). Manipulation of the circadian clock with benzodiazepines: Implications for altering the sleep-wake cycle. *Pharmacopsychiatry, 21*(1) 38-42.

Turek, F. W., & Van Reeth, O. (1989). Use of benzodiazepines to manipulate the circadian clock by regulating behavioral and endocrine rhythms. *Hormone Research, 31*(1-2), 59-65.

Van Der Linden, R. (1992). The American jet turns 50. *Aerospace America, 30*(11), 40-42.

Van Reeth, O. (1998). Sleep and circadian disturbances in shift work: strategies for their management. *Hormone Research, 49* (3-4), 158-62.

REFERENCES

Van Yperen, N. W., & Hagedoorn, M. (2003). Do high job demands increase intrinsic motivation or fatigue or both? The role of job control and social support. *Academy of Management Journal, 46*(3), 339-348.

Veillette, P. R. (2005). The facts on flight deck fatigue. *Business and Commercial Aviation, 96*(4), 46-49.

Waterhouse, J., Reilly, T. (1997) Jet-lag. *Lancet, 350*(9091), 1611-1617.

Waterhouse, J., Edwards, B., Nevill, A., Atkinson, G., Reilly, T., Davies, P., et al. (2000). Do subjective symptoms predict our perception of jet-lag? *Ergonomics, 43*(10), 1514-1527.

Winter, F. H., and Van Der Linden, F. R. (1992). Out of the past. *Aerospace America, 30*(10), 46-48.

Wright, D. G. (1947). Operational strain: stress in combat flyers. *Inter-Allied Conferences on War Medicine, 1*, 244-246.

Wright, J., Aldhous, M., Franey, C., English, J., & Arendt, J. (1986). The effects of exogenous melatonin on endocrine function in man. *Clinical Endocrinology, 24*(4), 375-382.

REFERENCES

Wright, N., Powell, D., McGowen, A., Broadbent, E., & Loft, P. (2005). Avoiding involuntary sleep during civil air operations: validation of a wrist-worn device. *Aviation, Space, and Environmental Medicine, 76*(9), 847-856.

Index

INDEX

About the Author

Bill Ragan, a native of Cleveland, Ohio, received a Bachelor's Degree in Psychology from Cleveland State University and a Master's Degree in Psychology from Walden University. He is a lifetime member of Psi Chi, the National Honor Society in Psychology, a member of the American Academy of Experts in Traumatic Stress, and affiliated with the United States Army Medical Department Regiment. Ragan's research interests include many aspects of clinical and aviation psychology.

Numerous concepts discussed in this book came from observations Ragan made when he worked in military aviation. His experiences and training on the Bell UH-1 Huey, the Bell AH-1 Cobra, and the Bell OH-58 Kiowa helicopters motivated him to pursue aviation psychology and this investigation into lag.

When he is not conducting research or writing about his findings, Ragan spends his time SCUBA diving in Florida. He also enjoys traveling, taking photographs, and spending time with his wife and family.

www.ingramcontent.com/pod-product-compliance
Lightning Source LLC
Chambersburg PA
CBHW022010170526
45157CB00003B/1222